I0468491

THE TRUTH ABOUT TRUTH

chains of evidence

By David Millett

Produced by

David Millett Publications
www.David-Millett-Publications.com

First published in the USA in 2016.
Second edition published in the USA in 2017.

ISBN-13: 978-1530563326
ISBN-10: 1530563321

Printed in the USA by Copyright © 2000 - 2016, CreateSpace, a DBA of
On-Demand Publishing, LLC.

Book production and design by David Millett.
Cover production and design by David Millett.

Contents

The Truth About Truth 1

Acknowledgments 1

Prologue.. 2

What is the nature of truth?.. 4

Hypothesis and theory...13

The source of truths 18

Can a story alone be truth?... 20

No truth is true forever...33

Why do we believe our truths? 35

What is human reality? 41

Mental illness and the supernatural 49

Bias or truth? 60

What is a lie? 63

The true nature of truth 66

Epilogue 70

About the Author 74

ACKNOWLEDGMENTS

I t is important to thank the people who helped me with this book. Without their support this endeavor would be much reduced.

For all the work she performed editing I give a special call out to Mary Sadtler. She absolutely kept me on the straight and narrow. Her insights were enlightening. Thank you, Mary.

Also, Laurence Chetwood spent time wading through my text and pointed me to some insightful ideas. He even inspired me to create a new chapter in the book. Thank you, Laurence.

A very special thanks to Dr. Julia Buss. This book would not have been started if it were not for the hours of conversation we have spent, over the last nine years, thinking about the notion of truth, the meaning of life, the universe, and everything. Thank you, Julia.

PROLOGUE

The human condition is fraught with ambiguity and plagued by uncertainty. We can't always know ourselves and what we might do in any given circumstance, even though we like to think otherwise. Therefore, we value the concept of truth as it is reassuring to feel that there is some kind of certainty in our world. If only we can find, define, and hold onto this elusive truth then we can soothe our psyches with the balm of truth, and thereby delude ourselves with feelings of certainty. It is not easy to accept that truth may be an outdated concept, or indeed a concept with very little utility, except perhaps in the realm of fairy tales and fantasy.

In our lives we can only see shadows on the wall of the human cave. We need to keep in mind these shadows are only built from our personal experiences, our culture, and our perception.

Defining truth is like trying to hit a moving target. If some idea becomes a so-called truth at some point, can it be an eternal truth? Are some truths immutable, or is this possibility mere wishful thinking? Is there a moment in time when circumstances allow a truth to be possible or to really be true? Then if that moment in time passes does the particular truth lose its relevance or use?

Often traditional truths are the most powerful in our cultures, and they are continually passed down through the generations. These types of truth gain immense hold over our lives and appear to gain extra power over us merely from their ancient lineage, regardless of their sense or nonsense.

Is it possible to have different versions of truth? Is a truth necessarily subjective and relative to situation? How much does truth matter to us, and in what ways does it control our decisions, even our lives. Does the popular concept of truth promote the accumulation of knowledge or hinder it?

This is a smart and insightful book that asks many such questions. It examines "truth" and questions assumptions about the idea of truth. It puts "truth" under close scrutiny and comes up with a useful tool for examining one's own, and society's assumed truths.

Dr. Julia Buss, 2016.

WHAT IS THE NATURE OF TRUTH?

I 'm telling you the truth!
This story is based on a true story.

Truth is stranger than fiction.

I swear to tell the truth, the whole truth, and nothing but the truth.

The infamous Donald Trump, said publicly:

> There were people that were cheering on the other side of New Jersey, where you have large Arab populations. They were cheering as the World Trade Center came down.

Was this true?

Kellyanne Conway, Counselor to the President of the United States of America said:

> Sean Spicer, our press secretary, gave alternative facts to that. But the point remains.

Can there be alternative facts?

There must be some reliable way to distinguish truth from fiction. How can we tell claimed truths from lies?

Should truth be absolute, or should it be flexible and change over time as new evidence is discovered?

We all toss the word truth around with abandon. I am just as guilty as the next person of using the word pliantly. But what does this enigmatic word really mean?

How should we use it?

When should we use it?

How do we know what we believe to be true is true?

I want to explore the true nature of truth. To learn what the truth really is about truth. It seems to me that the idea of truth is worthy of this effort.

Let's begin with the current Oxford Dictionary definition of the word truth:

Truth
Noun
1. The quality or state of being true.
2. That which is true or in accordance with fact or reality.
3. A fact or belief that is accepted as true.

Notice that the definition of the word truth has three main meanings. The first is self-referential (the quality or state of being true) and offers little insight into the nature of truth, so we can skip this definition.

The second meaning, I would suggest is the more widely held meaning of the word (that which is true or in accordance with fact or reality).

The third definition is the less understood version of this enigmatic idea (a fact or belief that is accepted as true).

The second definition incorporates the word "belief". Can a belief be the basis of a solid truth?

A belief is the acceptance that something exists or is true without proof. A belief is something one takes as true or real, such as a firmly held opinion. And finally, a belief is very often associated with a religious conviction.

Surely truth cannot be something based on fact or reality and belief at the same time?

The second and third definitions are therefore in contradiction to each other. The second aligns truth with facts and reality and the third associates truth with popularly held (cultural) ideas or beliefs. Can something be true just because many people believe it to be?

In this book I argue that truth can not be based on belief and that it must be based on facts and reality.

The concept of truth drives so many human behaviors (both good and bad) that it seems there is room for a clearer definition of this word.

Is truth based on fact or is something true just because many people believe it to be true?

In my opinion something defined as true must be based on facts and reality. Given this assertion the definition of truth put forward in this text is:

Truth is a complex and always changing concept. It is not absolute, binary, nor static. Useful truths should be available

to change as new evidence is found. Truth should not be
based in popular belief, but rather in facts and reality.

When using this definition of truth, then any truth that is claimed
to be static should be suspect. If a truth is claimed to be binary, yes
or no, black or white, true or false, absolute, then it is a questionable
truth. Because in our new definition of truth a reliable truth should
be able to change and grow over time.

The only absolute truth is that there are no absolute truths.

Truth based on absolutes cannot change, this is the first position
of systems that depend on absolute truth. If a truth cannot be
challenged and cannot be changed, then it is inflexible and
dogmatic. Inflexible truths have the effect of paralyzing thought and
reason.

Consider this example taken from the "New Living Translation"
of the New Testament, Mark 16:9:

After Jesus rose from the dead early on Sunday morning,
the first person who saw him was Mary Magdalene, the
woman from whom he had cast out seven demons.

That Jesus existed, was God, and transcended death is one of the
most popularly held absolute truths. Other than this story there has
never been another example in our collective human experience of
such an event. Yet this is one of the most popularly held absolute
truths.

If something cannot change, it cannot grow, adapt, or evolve then
it becomes frozen. Nothing else in life is like this. Everything about
us and around us changes over time, nothing is static or absolute.

The most destructive dimension of systems of absolute truths is
their ability to lockup or imprison human thought and reason. For
example, if we hold transcendence of death as an absolute truth

against all contradictory experience, facts, and data we find ourselves locked-up in a prison of unsubstantiated beliefs. Even with much evidence to the contrary we are imprisoned in this world, unable to escape it because the unbreakable bars of absolute truth keep us constrained. Believers in absolute truths must develop closed minds.

However, these seemingly unbreakable bars of absolute truth are restraints of our own creation. They are not physical constraints, only imaginary ones. We each of us hold the power to release ourselves from this prison of absolute truth. All we need do is reject these truths and seek truths based on chains of evidence.

The great bard himself (William Shakespeare) in his play Hamlet, famously said:

> There is nothing either good or bad but thinking makes it so.

Let us consider the following story as an example of the prison of absolute truths. Note religious thinking is not the only thinking that can be trapped in absolutes.

A man holds the absolute truth that obesity is due to people's freewill choices and nothing else. This man sees all overweight and obese people as weak-willed. His unwavering position begins to encourage hatred and rudeness towards overweight and obese people. He believes it is their lack of willpower alone that has caused their situation. His absolute truth ultimately precludes him from empathy, understanding, kindness, and more importantly rational thought towards the problems of overweight and obese people.

When this man is presented with chains of evidence that show obesity is also due to environmental, genetic, corporate, and other external influences he resists the new evidence in favor of his absolute truth.

Under the spell of absolute truth, this man is firmly locked up in a private prison of dogma. His reason is paralyzed, and his thoughts are frozen. It seems, nothing can change his mind not even new evidence.

Unwavering belief in the freewill of the individual made acceptance of additional information impossible to incorporate into this man's world.

Surely, this man is incorrect to reject new evidence? Does it not seem reasonable to reconsider his truth when new evidence is presented to him?

Are there any absolute truths at all?

The only one that I can think of is that you and I, all of us will die at some point. Is this the only absolute truth?

Consider, just for a moment, even this seemingly absolute truth may one day not be true. If science fiction imaginings of transferring human consciousness into machines comes to pass, then perhaps some might live forever. Could this be true?

If truth is not absolute, then what is it? Truth might be more like a scale. At one end of the scale truth is insubstantial or foggy and at the other end of the scale truth is substantial and perhaps even predictive, but never absolute.

Here is a scale defining this idea, our new concept of truth. I call it the truth-scale and offer it for your consideration:

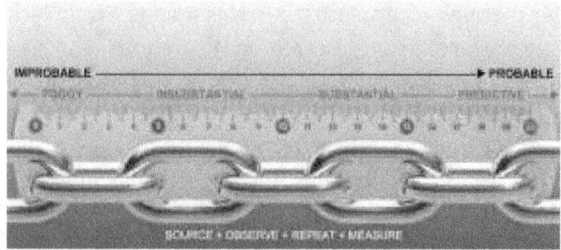

Notice that at the left side of the truth-scale claimed truths are improbable, foggy, or insubstantial. At the right side of the truth-scale claimed truths are highly probable, substantial, and predictive.

This is a more specific and useful definition of truth. Why? Because our newly defined truth-scale includes the notion of probability. What is the likelihood that a truth is true? Or what is the probability of a truth being true?

Probability is the measure of the likelihood that an event will occur. At the foggy end of our truth-scale the probability of a claimed truth of in fact being true is very, very unlikely. At the probable end of our truth-scale the probability of a claimed truth being true is highly likely.

In our new definition of truth, absolute truths are a clear sign there is an error in our thinking. However, keep in mind that when a claimed truth is placed at the foggy end of our truth-scale it is not completely untrue, rather it is highly improbable of being true. This dimension of our definition of truth is just as important as the idea that there are no absolute truths. Absolutely false ideas are just as useless as absolutely true ones.

This truth-scale based system of categorizing truths ensures that absolutizing does not become part of our concept for establishing truths, because we can never say something is absolutely true nor can we say something is absolutely false. This scheme of defining truth gives us the ability to place claimed truths on our scale of truth and assign a probability or likelihood a truth is true.

Why is it important for truths not to be absolute?

If a claimed truth is not absolutely true or absolutely false, then as new information is discovered the truth can be adjusted. When it is absolute, truth becomes dogma and unchangeable. Holders of absolute truths do not have open minds.

This idea of measuring truth via probabilities is not unprecedented. Probability is the concept behind one of the most powerful and impactful truths of the twentieth century: Quantum Theory.

Quantum theory is a branch of theoretical physics that strives to understand and predict the properties and behavior of atoms and subatomic particles. Without it, and the probability mathematics that drives it, we would not be able to build transistors, integrated circuits, or computers.

Erwin Schrodinger (Nobel Prize-winning Austrian physicist) came up with a famous thought experiment that extends the probability idea behind Quantum Theory from the very small scale, where it was first observed, to our normal human scale. His thought experiment gives us an insight into this bizarre idea, here is a summary of his thinking:

A cat is trapped in a box with a vial of poison that is released when a radioactive atom randomly decays. You cannot tell if the cat is alive or dead without opening the box. Schrodinger argued that until you open the box and look inside, the cat is neither alive nor dead but in an indeterminate state. This is because of the probability involved in the radioactive decay of the atom driving the experiment. The death or life of the cat is determined by randomness.

Hang on! How can a cat be both dead and alive at the same time? Surely this cannot be true? And yet, this is precisely the principles behind your cellular phone, computer, car, and just about everything we are reliant on in the twenty first century.

But if all truths are based on a probability of being true and not on absolutes, how do we place a truth on our truth-scale?

As it turns out humanity has developed, over many thousands of years, a very reliable way of establishing, identifying, and understanding truths. It is known as the scientific method and it's our best technique to distill truth. Why? Because it's first position is that any fact or truth is subject to change and modification based on new evidence. This ensures truths can grow, change, or be thrown out over time.

All other systems of establishing truth are based on absolutes or binaries. They are static, inflexible, and dogmatic. Religions for example use an absolute premise to establish truth. All religions espouse a position such as: there is a god or gods. And all religions will not allow the nonexistence of their god or gods no matter how much evidence is presented to the contrary. These claimed truths are therefore static or absolute.

But life is not static. From the day we are born we are changing. We grow in stature and sometimes wisdom. So why would truth be a static thing?

If new chains of evidence are so important to continually define, distill, and discover our truths, how can we ascertain reliable chains of evidence?

HYPOTHESIS AND THEORY

The best system humanity has created to identify truth is the scientific method. This grand method began with Aristotle (Greek philosopher and scientist) who first proposed the inductive-deductive method. Then Epicurus (Greek philosopher) who built on this idea and laid out his first rule for inquiry in physics. The scientific method was expanded by many others after this start, but it took on its modern persona after René Descartes (the French philosopher, mathematician, and scientist) wrote his famous work: "Discourse on Method".

The scientific method has a long history. During its long gestation it has been fired and tempered in the crucible of human thought and reason by many great minds. Here is a general description of this splendid method.

To begin a journey towards a truth in science a hypothesis is formed. Another way of understanding a hypothesis is to think of it as an idea of something that might lead to a truth. Next experiments are designed and then conducted to disprove the hypothesis. In the experiment observability, repeatability, and measurement are sought. If these experiments cannot disprove the hypothesis we move a step closer to a theory or a truth.

In science all theories and all hypotheses are always subject to change as additional information is discovered. This is the power of the scientific method.

For example, today Albert Einstein's truth that light will bend due to gravitational effects is one of the most observed and established scientific truths. However, when Einstein first came upon this truth he had only one of the three elements of the scientific method: measurement. He had mathematics that proved his idea, but at that time he lacked any observation let alone repeated observations.

Many years went by before anyone attempted to observe his predictions. After a failed first attempt an observation was finally made that did indeed verify his forecast.

In 1919, during a total solar eclipse of the sun, Sir Arthur Eddington performed the first successful experimental test of Albert Einstein's truth that: light will bend due to gravitational effects. Then, after this initial observation, every other has verified that light does indeed bend due to gravitational effects. In the end Einstein's great truth took on all three elements of a truth worthy of a prominent position on our truth-scale: observability, repeatability, and measurement. This chain of evidence distills and refines all truths.

Another way of expressing this is that the bending of light due to gravitational effects is highly probable. This places Einstein's great truth high on our truth-scale.

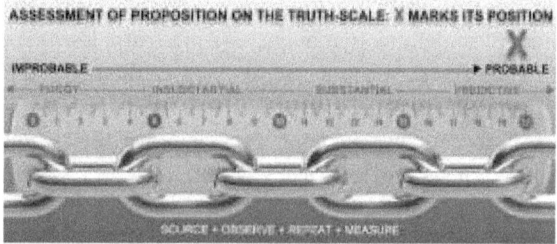

An additional indicator of a substantial truth is that it can make accurate predictions about our universe.

Albert Einstein formulated many other truths during his lifetime. These truths are so distilled and refined that they too can make accurate predictions. For example, Einstein's first theory of relativity predicted the possibility of black holes back in 1916. The term "black hole" was coined in 1967 by the skeptical American astronomer John Wheeler, and the first one was discovered in 1971.

Another example of a highly distilled and refined truth is Charles Darwin's grand theory of evolution by natural selection. This truth also contains all three elements of a substantial truth: observation, repeatability, and measurement.

Just as with Einstein's truth, Darwin's truth has reached the level of accurate prediction of the universe. The discovery of the Darwin moth (Xanthopan morganii or Morgan's sphinx moth) is one example of the predictive power of Darwin's truth.

The great scientist postulated the existence of this unknown species of insect based purely on the size of an orchid from Madagascar (Angraecum sesquipedale) and his understanding of the evolution and the ecology of orchids and insects. Darwin postulated that there must exist a highly adapted insect, with just the right length proboscis, that fertilizes this special orchid. It wasn't until 1992, nearly a century after Darwin's prediction that observations were made of the elusive moth feeding on the flower and transferring pollen from plant to plant. This was observed in the wild and confirmed further with studies in captivity.

Darwin's truth like Einstein's truths continue to make accurate predictions of the universe. This dimension of both truths assures them an elevated position on our truth-scale.

Another example of the predictive powers of Darwin's grand truth is based on the idea of common ancestry. Common ancestry is a major principal of Charles Darwin's grand theory of evolution by natural selection. It leads logically to powerful and testable predictions about evolution. For example, if we see that birds and reptiles group together based on their features and DNA sequences, we can predict that we should find common ancestors of birds and reptiles in the fossil record. Such predictions have been fulfilled, moving the truth of evolution by natural selection even higher on our truth-scale.

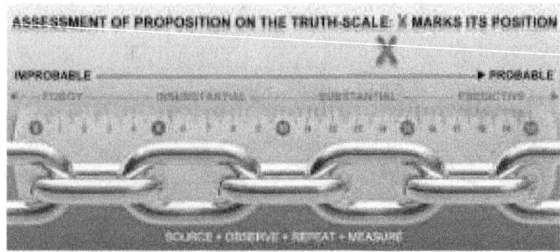

If these predictions were not enough, one of the greatest fulfilled predictions of Darwin's great truth was the discovery, in 2004, of a transitional form between fish and amphibians. This is the fossil species Tiktaalik roseae, which tells us a lot about how vertebrates came to live on land.

Until about 390 million years ago the only vertebrates were fish. But, 30 million years later, we find (in the fossil record) creatures that are clearly tetrapods or four-footed vertebrates that walked on land. These early tetrapods were like modern amphibians in several ways, they had flat heads and bodies, a distinct neck, and well-developed legs, and limb girdles. Yet they also show strong links with earlier fishes, particularly the group known as "lobe-finned fishes", so called because of their large bony fins that enabled them to prop themselves up on the bottom of shallow lakes or streams.

The fishlike structures of early tetrapods include scales, limb bones, and head bones. Tiktaalik roseae's discovery (in the fossil record) is a stunning example of the predictive power of the theory of evolution.

Where does Charles Darwin's great truth of evolution by natural selection finally fall on our truth-scale? It is a highly probable truth.

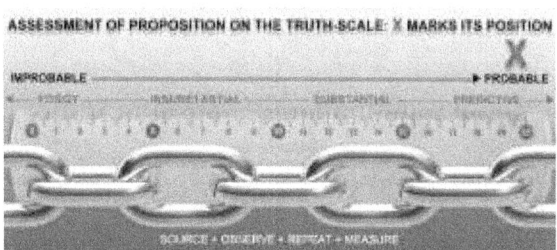

Any truth worthy of an exalted position on our truth-scale must be based on a chain of evidence that is: observable, repeatable, and measurable. If any one of these elements is missing from a claimed truth, then it must be considered suspect and placed firmly at the foggy end of our truth-scale.

THE SOURCE OF TRUTHS

When considering truths, one must also carefully examine the source of the claimed truth. In the case of Einstein's truth of the bending of light and Darwin's truth of evolution by natural selection the source of these truths were two independent scientists with a passion for the collection of knowledge. They were not supported by any industry nor any particular social influence. Because of this and their collective body of other works the source of these two great truths is very reliable.

Also, the credibility of these two sources is increased by the repeated independent verification of both of their truths over the years. Many other scientists have verified these truths independently. This community of scientists is an authoritative group and one that is extremely competitively self-policing. Being open in communicating findings and competitive in establishing the validity of ideas is one of the most powerful systems behind modern science and in complete contrast to other schemes of establishing truth.

Systems of establishing truths that are fixed and based on dogma typically have very poor sources behind their claims. For example, Muslims believe that Islam is the complete and universal version of a primordial truth that was revealed many times before via prophets including Adam, Noah, Abraham, Moses, Jesus, and Muhammad. But, these claimed sources of their truths are primarily revealed to us via the very scriptures that cite these same sources. To say this is circular is an understatement.

The chains of evidence that Adam, Noah, Abraham, Moses, Jesus, and Muhammad existed beyond the pages of these scriptures is very slim. As for the text they claim is the source of their truths (the Qur'an), Muslims consider it to be both the unaltered and the final revelation of their god. It is fixed, unchangeable, unverifiable, and dogmatic.

A well-established, trustworthy truth has a reliable source and contains the three essential elements of all dependable truths: observability, repeatability, and measurement. If a truth consists of these elements, it often has the added ability to make accurate predictions about our universe. These elements of a good truth can be referred to as the chain of evidence. Like all strong chains each link must be unbroken and secure. If one is weak the chain is very likely to break leaving the claimed truth stranded at the foggy end of our truth-scale.

The application of the scientific method, or performing double blind experiments, might not be practical to every truth we analyze in daily life. However, before we consider a claimed truth looking to its source and the motives driving this source will certainly be time well spent. We should also try and introduce logic, skepticism, and scientific curiosity to our assessment of claimed truths. Consideration of a claimed truth's feasibility and usefulness will also help us better position truths on our truth-scale.

By David Millett

CAN A STORY ALONE BE TRUTH?

We humans have developed many ways to communicate ideas. One of the oldest forms is story telling. Way before we could write we could speak and we soon learned to use this ability to create stories. Human stories have so often led to claims of truth, but can a truth be based on a story alone?

Sigmund Freud presented the idea that most of our mental life, including most of our emotional life, is unconscious at any given moment; only a small component is conscious. Given this base nature of the human mind, in the following story I would like to introduce two concepts. The first is that we humans are driven by our emotions first and our reason second. It is this fact alone that accounts for most of the unclear thinking and foggy truths that exist in the world today.

The importance of emotions to our decision making and therefore reason was tested a few years ago. Antonio Damasio (Professor of Neuroscience at the University of Southern California and an Adjunct Professor at the Salk Institute) made a groundbreaking discovery. He studied people with damage in the part of the brain where emotions are generated.

He found that they seemed normal, except that they were not able to feel emotions, and they all had something peculiar in common: they could not make decisions. They could logically describe what they should be doing, yet they found it very difficult to make even simple decisions. For example, they could not decide what to eat. Many decisions have pros and cons such as: shall I have the chicken or the turkey? With their emotions disabled and no clear rational way to decide, these test subjects were unable to arrive at a decision.

Clearly emotions are a critical part in our ability to make decisions. In fact, in evolutionary terms one can see that emotional decisions give us an advantage, they allow us to make fast decisions. These decisions may not be well reasoned, but sometime faster is better than well-reasoned especially when you are about to be eaten by a lion.

The second point I wish to highlight in the following story is that well-reasoned truths must contain all three elements of our chains of evidence (observation, repeatability, and measurement) to be given an elevated position on our truth-scale.

The following story is a fictional application of emotions driving the establishment of truth.

The story of the crocodile

If a man saw his brother taken by a crocodile at a particular river, he would no doubt remember this event. The man's emotions would be driven to bursting after witnessing such an event. The death of his brother would be indelibly imprinted on his mind. We can all relate to this.

Later in the man's life when he became a father and he wished to convince his daughter not to go too close to the same river for fear of a crocodile eating her, he might try and tell her directly about the tragic event he had witnessed years ago.

Children being children the father's initial story of the tragic event might not be impactive enough to ensure she did not go near the river. If this direct method did not impress the girl, he might resort to embellishing the story.

Soon his story becomes a tale about a terrible monster that lives in the river. He would no doubt discover that his new fable was more effective if he exaggerated it and overstated the threat. The crocodile might become gigantic. It could take on hypnotic powers. Its claws and teeth might grow longer, and its speed of motion become supernatural. If he could impress the girl with the new and embellished story he would be sure she would not suffer the fate of his brother. This is in fact all the father cares about.

What is the truth here?

It is that there is a potential of attack on the girl by a crocodile. The father knows this threat to be true as he saw his brother taken by a crocodile with his own eyes.

For a thing to be a reliable truth it must have a solid chain of evidence that supports it. This chain of evidence consists of three things: observability, repeatability, and measurement. As well as these three essential elements it is important to gauge the reliability of the source of a truth.

The girl's father is an authoritative source from the girl's perspective. The father in our example has also established one of the three elements that support a reliable truth. He saw (observed) his brother taken by a crocodile at a particular river. So, his truth is almost true because his truth consists of direct observation. Is this right?

No not really.

It is at this point that we must consider how emotions are involved in our story. The father cares little about empirical evidence to support his truth. To him his observation and his emotional shock is all he needs to support his perceived truth that the river is infested with man-eating crocodiles.

The man firmly believes there are crocodiles in the river. But, what is the truth about this river?

Consider, during the years since his brother's death as the girl grew, have there been other attacks at the river? If the father had witnessed other attacks over the years, then his truth would have taken on the second element of a probable truth: repeatability.

Now for his truth to reach up to the levels of a very reliable truth our story-father would need to count the instances of crocodile attacks at the river over the years. This would add our third element of a substantial truth: measurement.

With all three elements in place his truth might even take on predictive powers. But, this particular truth does not contain all three elements and therefore falls far short of being true. It must remain just a story, and must be placed at the foggy end of our truth-scale. That is to say it is highly improbable, that there are still crocodiles in this river.

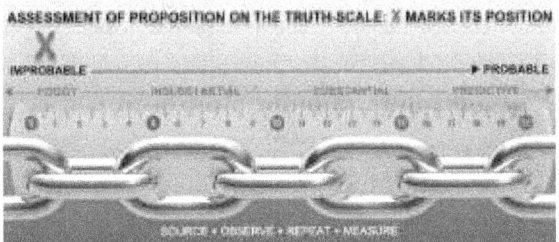

But surely all this particular father cares about in regard to this truth is that his daughter does not get eaten by a crocodile. He finds no need to seek repeatability nor find measurable data to support his truth; as his story worked to achieve his goal and kept his daughter away from the river.

But are there really any crocodiles left in the river to eat his daughter? And what of the embellished abilities of the crocodiles in his story?

This is one of the dangers of elevating stories to the level of truths. The result of the father's story is that these imaginings have been firmly imprinted on the girl's mind. Two things have happened to the girl because of her father's story. She now holds as truth that crocodiles live in the river and that these crocodiles have supernatural abilities. But, without establishing all the elements of a reliable truth we cannot know the truth about crocodiles in this river.

When Albert Einstein first revealed his claim that light bends under the influence of gravity. He used mathematics to prove his truth. But, even with mathematical proof, scientists of the time found his idea preposterous. It was not until much later, when experiments conducted during total eclipses of the sun verified his mathematical prediction, did scientists begin to realize Einstein's truth was indeed true. When the experiment was repeated many times with the same result the truth of Einstein's predicted bending of light due to gravitational effects was established.

At first Einstein had provided measurable evidence only of his truth, but without observation and finally repeatability even this great truth was considered false.

Remember that one of the definitions of truth is: <u>That which is true or in accordance with fact or reality</u>. But, what is reality? Should reality be based on stories alone?

Our crocodile in the crocodile story took on supernatural powers as the father in the story began to embellish his observation to ensure observance by his daughter of his wish for her not to go near the river.

Truths are very often the building blocks of human reality. But as we have discovered stories are not good chains of evidence for truths. Therefore, stories alone are not necessarily good foundations for human reality.

Let us for the sake of understanding examine a widely held claimed truth, which is based on the Judeo, Christian, Islamic, creation story.

> In the beginning God created the heavens and the earth. Now the earth was formless and empty, darkness was over the surface of the deep, and the Spirit of God was hovering over the waters.
> And God said, "Let there be light," and there was light. God saw that the light was good, and he separated the light from the darkness. God called the light "day," and the darkness he called "night." And there was evening, and there was morning—the first day.

We will skip to the last paragraph of this well-known story before we all fall asleep. If you feel compelled to read the complete tale, I'm sure you'll be able to find a copy.

> God saw all that he had made, and it was very good. And there was evening, and there was morning—the sixth day.

This story is claimed by many to be a literal and absolute truth. The belief in this tale as a truth leeds to countless modern antisocial behaviors and much suffering in the world.

Why?

Because belief in this story leads to the certainty that many other bible stories are also true. Many of these stories promote prejudice and injustice. Given the dangers of holding bible stories as true it is worth us spending some time testing this story.

Let us examine this creation story and see how it holds up to our criteria for a reliable truth. Let's find a place for this creation story on our truth-scale.

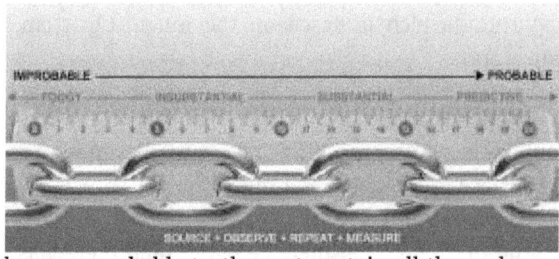

As we know, a probable truth must contain all three elements of the chains of evidence: observability, repeatability, and measurement. We also know we must consider the source of a claimed truth.

What is the source of this story? It is claimed by its supporters to have come directly from their god. Has there ever been any direct observation of this claim? No, there has never been any observable, repeatable, or measurable evidence that a god wrote this text. The only support of a god writing this story is the story itself and its supporters, which seem overloaded with bias.

The words the text is based on come from ancient tablets and manuscripts this is so. However, once again there is no chain of evidence that links the creation of these ancient tablets and manuscripts to a god. In fact, the only observable, repeatable, or measurable evidence leeds us to see that these ancient tablets and manuscripts were written by men and not gods. So, the source of this truth is suspect and pushes this truth firmly towards the foggy end of our truth-scale.

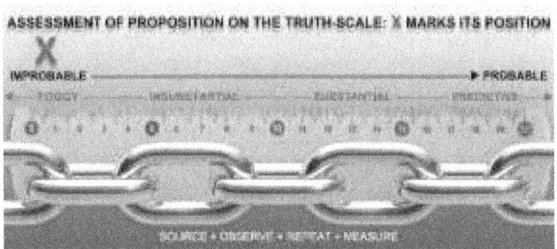

We have just read the creation story and its claim to be a truth. We do observe directly the world that it claims was created in the story. Does this meet our requirement of observability?

Not really.

We have never observed the genesis process as defined in this story on the Earth or any other planet in our current experience. Any evidence that meets our required level of truth does not seem to confirm this story. Observations made by geologists, biologists, astronomers, and paleontologists (in fact every field of science) show clearly the Earth has been around for at least four billion years.

The forces of plate tectonics, volcanism, wind, rain, and sun shaped the surface of the Earth. Chains of evidence show that life started as simple single celled creatures and via Darwin's grand truth evolved into the more complex multi-celled creatures of today.

Observable, repeatable, and measurable evidence leeds us to see the universe itself to be 13.5 billion years old, not six days as the story outlines. This lack of observation of the genesis process moves this popular story towards the foggy end of our truth-scale.

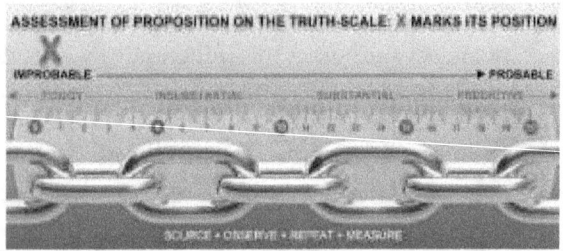

The supporters of the story do insist their god communicated this tale directly to the authors. Has this type of direct communication from a god ever happened again?

Other religions since and before the Judeo-Christian-Islamic religions have made similar claims of direct communication from a god, but have supplied the same level of evidence of such a communication.

So, such god communication has been claimed to have been repeated, but with the same lack of evidence that such a communication ever emanated from a god. This lack of evidence has been consistent over repeated claims of communication from gods.

This consistent lack of a chain of evidence keeps this claimed truth firmly at the foggy end of our truth-scale.

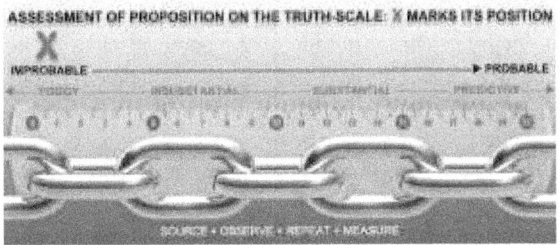

Is there any way to measure this claimed truth of a god or gods writing this story?

Over the years many experiments have been formulated and tested to prove that external gods or other supernatural forces exist, but none has stood the test of time. This lack of a chain of evidence to support measurement of god communication firmly strands this bible story at the very, very foggy end of our truth-scale.

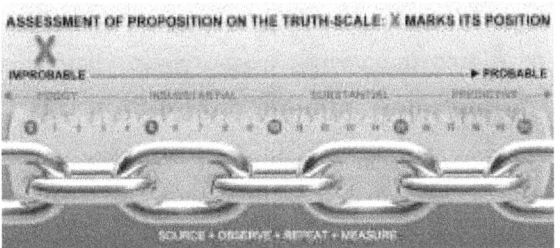

But, so many people believe this story to be true surely that makes it true?

Not at all.

The definition of truth we have been constructing in this text is not a matter of wishful thinking or widespread belief. Just because a lot of people wish or believe something to be true does not make it so.

When Albert Einstein claimed that light was not fixed, but subject to bending by gravitational forces the popular scientific position of the time was that light was fixed and could travel only in straight lines. The widely held belief was incorrect as it turned out.

If a truth is claimed solely on a story that cannot be confirmed by our truth tests of observability, repeatability, and measurement, then these claimed truths must be placed very low on our truth-scale. In the case of this bible story it does not meet our standard of truth it's just a popularly held, very old, and highly improbable story.

Yet there are many people who would die or kill if you were to contradict this story. How can we defend against such adamant believers?

Given our newly defined concept of truth we cannot ever definitively claim that a popularly held truth is not true, even when it has no evidence to support it. We can only say that it is very, very improbable. You might think that this is a flaw in our system of truth as it leaves open the argument by, foggy-truth supporters, that we can never disprove their claimed truth; so therefore, their foggy-truth is true and deserves an elevated position on our truth-scale.

Fortunately, there was once a very clear thinker: Bertrand Russell (British philosopher, logician, mathematician, historian, writer, social critic, and political activist). The great philosopher proposed the sobering idea that the philosophic burden of proof lies upon a person making scientifically unfalsifiable claims rather than shifting the burden of proof to others.

Said in another way: If a claimed truth does not contain the chains of evidence we now require for a highly probable truth, then it is up to the claimer of this truth to meet our burden of proof. It is not up to us to make their case for them.

His postulate is known as Russell's teapot, sometimes called the celestial teapot, or cosmic teapot. It got its name from his argument:

> If he claims that a teapot orbits the Sun somewhere in space between the Earth and Mars, it is nonsensical for him to expect others to believe him on the ground that they cannot prove him wrong.

Because he claims there is a teapot orbiting Mars and because we cannot disprove his claim, his claim must be true.

Can this be so?

No. It is highly improbable that his claim of an orbiting teapot is true, even if we cannot disprove this claim directly. And so, his claimed truth must fall to the lowest point on our truth-scale.

And yet this is exactly how all religions support their truths. Rational religious believers may concede that there is little evidence to support their claimed truth and, yet they hold it up as a truth on the basis that we cannot definitively disprove it. However, the burden of proof of their claimed truth is theirs not ours. It is up to them to supply us with a reliable source and a clear chain of evidence of their claimed truth if it is to be moved up our truth-scale.

Where on our truth-scale does our proposition fall that: Stories alone are not good foundations for human reality. This is a highly probable truth.

ASSESSMENT OF PROPOSITION ON THE TRUTH-SCALE: X MARKS ITS POSITION

IMPROBABLE ————————————————→ PROBABLE

FOGGY · INSUBSTANTIAL · SUBSTANTIAL · PREDICTIVE

SOURCE + OBSERVE + REPEAT + MEASURE

NO TRUTH IS TRUE FOREVER

Given where we have traveled so far on our journey to understand truth, it is important to appreciate that every truth will eventually be, at least improved, if not completely replaced given enough time. This is so for even our greatest truths of the modern era, in physics and biology.

For example, Albert Einstein's theories of relativity have recently had to cope with what might one day supersede these powerful truths.

In 1995 Edward Witten (American theoretical physicist and professor of mathematical physics at the Institute for Advanced Study in Princeton) started what has been called the Second Superstring Revolution by introducing M-theory to the world.

This theory combines the five different string theories (along with a previously abandoned attempt to unify General Relativity and Quantum Mechanics called D-Supergravity) into one theory.

At the moment M-theory suffers from the very same issue Einstein faced when he first presented his grand theories; it does not have all three elements of a probable truth as yet. M-theory is expressed in mathematics at the moment and so supports only one of the three ingredients of a reliable truth, which as we know so well is observation, repeatability, and measurement. However, if the past is to be repeated then in time M-theory will gain all three elements or it will fall away to be replaced by some yet unknown insight into the physical world.

This is the nature of truth. It is not static, binary, or absolute. It is flexible, ever changing, and growing.

Another example of this dimension of truth is Darwin's grand theory of evolution by natural selection. In recent years, a new line of research known as epigenetics has challenged some of the reasoning behind evolution by natural selection.

In the field of genetics epigenetics is the study of cellular and physiological phenotypic trait variations. It is an explanation for how creatures express their physiology in the world.

The theory of epigenetics claims that these variations are caused by external or environmental factors, importantly, during the creature's lifetime. It is hypothesized that these external or environmental factors switch genes on and off and therefore affect how cells read these genes. This process in turn affects how organisms express their physical traits. This contrasts with the idea behind evolution by natural selection that supposes these changes are only caused by direct variations in the DNA sequence of creatures, then these changes are expressed in subsequent generations.

Epigenetic theory exposes a potential modification to the current ideas behind Darwin's grand truth of evolution by natural selection. But, this type of forging and purifying of truth is part of the nature of truth we have defined in this text.

This process of distilling truth is in fact the very thing that makes our system of truth stronger than other systems. Truths based on absolutes typically lock up and imprison our reasoning. Our new system sets our reasoning free.

WHY DO WE BELIEVE OUR TRUTHS?

To answer this complex question let us once again consider the girl in our crocodile story.

Her father was so concerned that she would be eaten by a crocodile at the river that he invented a powerful story, based on a personal and emotional observation. He embellished his story to impress her and ensure she would never venture to the river. The result of this story was the girl believing in an unreliable and unsubstantiated story. This weak truth became part of the girl's reality.

Her father, in her eyes, is an authority and so eventually the story became truth or reality for the girl, she believes sincerely that the river is overrun with crocodiles that possess supernatural powers.

There are many reasons why we believe a story to be truth. Some reasons are based in our chain of evidence and some not. In the case of our crocodile story the father had only good intentions in mind when he embellished his story and told it to his daughter. He did not want her to be eaten by a crocodile.

As the daughter took on the story as being true, for her the river truly was infested with monstrous supernatural crocodile creatures. They may only live in her imagination, but that does not make her truth any less true to her. But, the father's intention was to protect the daughter not to warp her sense of reality. And yet this is exactly what he did.

This phenomenon highlights another dimension of truth, that is it can be very personal or subjective. It might not hold a prominent position on our truth-scale, as the chains of evidence supporting it are lacking, but we will still hold it to be true.

The crocodile story also highlights how we can give up responsibility for discovering or refining the truth. The girl in the story deferred her reality to her father's story. She did not bother to verify it.

In 1963, Stanley Milgram submitted the results of his obedience experiments in the article "Behavioral Study of Obedience". Milgram conducted experiments on human obedience to malevolent authority. His work became the most important social psychological research done in his generation.

Milgram stated in his study report:

> The essence of obedience consists in the fact that a person comes to view himself as the instrument for carrying out another person's wishes, and he therefore no longer regards himself as responsible for his actions.
>
> Once this critical shift of viewpoint has occurred in the person, all of the essential features of obedience follow. The adjustment of thought, the freedom to engage in cruel behavior, and the types of justification experienced by the person are essentially similar whether they occur in a psychological laboratory or the control room of an Intercontinental Ballistic Missile (ICBM) site.

Milgram conducted his experiment to try and explain the behavior of German citizens during the years of the Nazi's Third Reich.

He recruited a subset of people from the entire New Haven community of (then) 300,000 people. He found that out of 296 participants 65 percent of study participants would continue to administer (what they believed to be) electrical shocks to someone in another room. This high percentage of people would continue to administer, supposed, electric shocks to and beyond the point they suspected the receiver of the shocks was unconscious or worse.

All that was needed to have 65 percent of study participants conduct this inhuman act was to ask them to do it in the name of the experiment.

Importantly, when he much later interviewed participants that administered the presumed shocks many of them claimed they would never perform such an inhumane act; and yet they had done so.

We can all believe things to be true about ourselves, our world, our reality that are not so. Milgram's experiment shows this clearly.

Imagine we conducted a survey of our crocodile girl's river over several years. For the sake of discussion let's assume we recorded no further crocodile attacks or even sightings during this period of repeated observations. We could say with an important level of probability that crocodiles no longer frequent this river.

We explain our finding to the girl, but she firmly holds on to her truth. Our facts do little to dissuade her. Her truth has become part of her reality, she has become invested in the claimed truth. It becomes very difficult, if not impossible for her to consider another truth, another reality.

The same can be said of those people in the twenty first century who still hold the Judeo-Christian-Islamic creation story to be true. The story was most likely told to these believers at an early age and claimed to be true by authoritative sources such as parents or religious community figures. Just as with our crocodile girl these believers of the creation story hold it up as a truth, it forms their reality. And just like the girl in our story their truth is very true to them.

When we present facts, to these believers that contradict their story they are rejected. Just like the girl in our story is not convinced by facts that contradict her truth.

Surely, a truth's relevance and utility must change as circumstances change over time. All truths become anachronistic eventually. If our minds are closed, how would we know?

An acquaintance of mine once said this to me:

"I was born a Catholic and will die a Catholic."

I said, "What evidence can I present to you that might change your mind about the existence of your god?"

He responded, "Nothing will ever change my mind. I have faith and faith is all I need."

"This seems closed-minded to me," I replied.

"I have a very open mind," he retorted.

Can you have an open mind if you are not prepared to change your view based on new evidence? Or based on the discovery that the truth you hold to be true is lacking in a chain of evidence that supports it.

No, it is not possible to claim an open mind if you are unwilling to change your mind based on new evidence.

Changing truth based on new evidence is the cornerstone of the scientific method and our proposed system of defining truth here in this text. However, changing truth based on new evidence is antithetical to all religions.

Inflexible thinking is an artifact of religious belief.

Why?

Because absolute truths lead to the rejection of evidence that might contradict the claimed absolute truth.

Consider for a moment the shameful and abhorrent actions taken against Galileo Galilei (Italian astronomer, physicist, engineer, philosopher, and mathematician) by the Catholic Church in the sixteenth century.

Galileo was threatened with execution for his well-reasoned truth that the universe was centered around the sun and not around the Earth. The doctrine of the Catholic Church at that time held the absolute truth that the Earth was the center of the universe.

Galileo had established observable, repeatable, and measurable evidence of a sun-centered universe. He was at first not prepared to withdraw his claim. The church working in its system of absolutes was unconvinced by Galileo's chain of evidence and so threatened to excommunicate, torture, and kill the great scientist.

Galileo was ultimately not executed, but rather placed under house arrest for the rest of his life and forced to publicly recant his truth.

Today, via our chain of evidence we know Galileo was more correct that the Sun is the center of our universe and the Catholic Church was wrong about this truth.

I must mention here that even Galileo was not quite correct. Since his postulate more evidence has shown the Sun to be at the center of our solar system and not our universe.

But surely, the Catholic Church is all about truth? How could this terrible injustice happen? It happened because the Catholic Church uses a flawed system of absolute truths. This system is prone to fixed, unreasoning, and closed-minded thinking. This is what you have signed up for if you associate yourself with such institutions as Christianity, Judaism, or Islam.

The Catholic Church did recently change its official opinion on an Earth Centered universe to Galileo's position. Of course, it took them nearly 500 years to escape the prison of their closed-minded thinking, which proves how debilitating absolute truths can be.

A better system of searching for, and establishing truth is one based on the idea that truth is flexible and open to change over time as new chains of evidence are found. This is the basis of the scientific method and therefore it is a superior system for seeking out and revealing truths.

Where then, on our truth-scale, does the proposition that: Absolute truth is a flawed system for defining truths fall. This is a highly probable proposition.

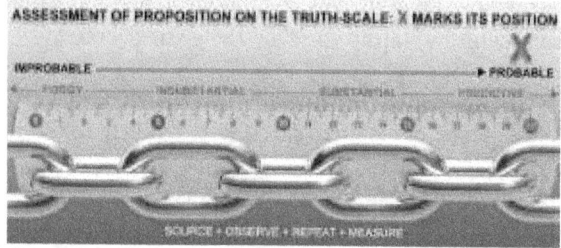

The girl's truth, in our story about supernatural crocodiles and the truth believed by Judeo-Christian-Islamic creation story adherents are nonetheless real to these believers. Which leads us to the question, what is real?

WHAT IS HUMAN REALITY?

Let us for the sake of understanding consider the following idea for a moment:

Human reality is derived. It is personal. It is an invented construct made up of input from our senses that is fed into our nervous system, and continually processed by comparing this ever-streaming input into and through our emotions, memories, experiences, and lastly our reason. Therefore, human reality is not separate from our physiology; it is in fact our physiology.

Can this be true?

Consider, if an external event is perceived by two people usually it coincides with external reality, but there is a chance that they may perceive or remember the event differently.

Why?

Because each person has a differently formed nervous system. Each has a distinct set of emotions, memories, experiences, and reason.

Human reality begins and ends at the boundaries of our nervous systems. In an effort to establish the truth of this idea let us consider that human reality can be intentionally influenced as the following example shows.

Elizabeth Loftus (American cognitive psychologist and expert on human memory) performed experiments in the mid-seventies demonstrating the effect of a third party's introducing false facts into subject's memories.

During her experiment subjects were shown a photographic image of a car at an intersection with either a yield sign or a stop sign. Experimenters asked participants questions, falsely introducing the term "stop sign" into the question instead of referring to the yield sign participants had actually seen.

Similarly, experimenters falsely substituted the term "yield sign" in questions directed to other participants who had actually seen a stop sign image. The results indicated that the probability of subjects recalling seeing the false "introduced" images increased.

In the initial part of the experiment, subjects also viewed a photographic image showing a car accident. Some subjects were later asked how fast the cars were traveling when they "hit" each other, others were asked how fast the cars were traveling when they "smashed" into each other.

Those subjects questioned using the word "smashed" were more likely to report having seen broken glass in the original image, which did not exist. The experiment established that the introduction of false cues altered participants' memories.

Clearly, this experiment shows that human memory, which is a key component of human reality, can be influenced by language and images. This alludes to the idea that human interpretation of reality is personal.

This personal human interpretation of reality is sometimes a very low fidelity version of objective external reality.

Let us consider, this additional example of the proposition that human reality is personal and does not extend past our nervous systems:

To treat refractory epilepsy (epileptic seizures that can vary from brief and nearly undetectable to extended periods of vigorous shaking) an operation has been performed by neurosurgeons that cuts the corpus callosum within the patient's brain.

By severing this part of the human nervous system, the result is a split-brain. That is because this physical brain structure connects the left and right hemispheres of the brain together. Once cut, communication between the two hemispheres is no longer possible. After the right and left brain are separated, each half will have its own separate perception, concepts, and impulses to act.

Having two "brains" in one body can create some interesting dilemmas. For example, when one split-brain patient dressed himself, he sometimes pulled his pants up with one hand (that side of his brain wanted to get dressed) and down with the other (that side didn't). It was reported that once this patient grabbed his wife with his left hand and shook her violently. So, his right hand came to her aid and tried to restrain his aggressive left hand.

A popular truth is that our perception of reality is in total sync with external reality. Yet when the brain is damaged we perceive reality differently.

Both these examples show human reality to be much less fixed and much more malleable and personal than generally understood. A fixed shared human reality is a widely held truth, but these chains of evidence show this belief to be incorrect.

Consider for a moment, if our proposed idea that human reality is constructed due to input from our senses that is fed into and processed by our nervous system. Then the observed, repeated, and measured behaviors of participants in Elizabeth Loftus' experiments make perfect sense.

Why?

Because our proposed model of human reality is subject to personal interpretation of external events. It is not limited to an absolute universally shared reality. Human beings are not machines capturing with perfect fidelity a universal, fixed, and common external reality.

However, if human reality is based on the idea of a fixed distinct external reality, common to all, and that reality is separate from our physiology, then Elizabeth Loftus' experimental results are harder to reconcile.

The popular idea of a fixed distinct commonly-shared external reality moves closer to the foggy end of our truth-scale because of the results of Elizabeth Loftus' experiments.

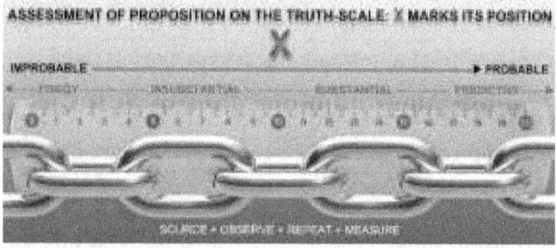

Because our proposed idea of human reality is based on the concept that it is physically constrained and held in our nervous system. This leads to the thought that when we die, and our nervous system decomposes into dust and detritus; personal human reality stops.

But, intentionally missing from this understanding of human reality is the popularly held truth of a soul, or spirit, or everlasting life-force.

Which of these ideas is true then?

Is it true that, human reality is constrained by the nervous system or is it true that an immortal soul exists?

Surely the popular idea that we all possess an immortal soul must be true? To discover the truth here let us use our newly found system of establishing truth to find an answer.

If the idea of a soul, or spirit, or everlasting life-force is true then the change in behavior of the unfortunate recipients of the split-brain procedure outlined above makes little sense.

Surely, if there is one human soul, which is separate from human physiology then disturbances in human brain structure would not interrupt this supposedly distinct entity?

Our claimed truth that human reality is limited to our nervous system does explain the observed, repeated, and measured behavior manifest by split-brain procedure patients.

However, the notion of a soul, or spirit, or everlasting life-force that is separate from the nervous system does not predict the kind of behavior displayed by split-brain patients. This supernatural idea of an everlasting life-force begins to slip down our truth-scale towards the foggy end.

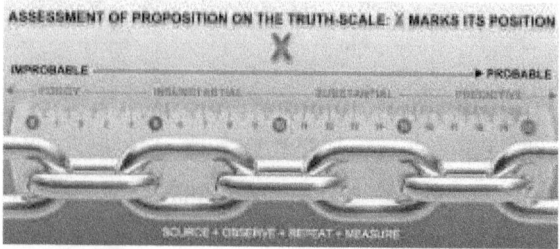

What is good evidence?

Observable, repeatable, and measurable events make for the most reliable evidence. Events based on this chain of evidence may be confidently placed high on our truth-scale and will often make reliable predictions of the external universe.

What is bad evidence?

Hearsay, old scriptures, wishful thinking, information passed on via culture, popular ideas, and even eyewitness reports that cannot be verified are all examples of bad evidence. Truths based on this line of thinking must sit low on our truth-scale and will typically not make any reliable and accurate predictions of our universe.

A soul, or spirit, or everlasting life-force is a popularly held truth. It is also supported by most religions in one form or another. This idea is also a truth that is taught to many children from an early age by authoritative figures in these children's lives. And so, this idea has become a truth held by most humans on the planet today.

But what is the chain of evidence that supports this widely held truth?

As it turns out the evidence is slim. The existence of separate human spirits has been tested on more than one occasion by reputable, and not so reputable people. On every occasion the results have been less than conclusive. Repeated experiments over the years have resulted in the same lack of observable and repeatable evidence. Extensive attempts to quantify this widely held truth have also fallen short.

We've already established that just because an idea is widely held or popular does not ensure it an elevated position on our truth-scale. So, we can put this idea's popularity aside with confidence. This claimed truth of a soul, or spirit, or everlasting life-force slips deservedly towards the foggy end of our truth-scale.

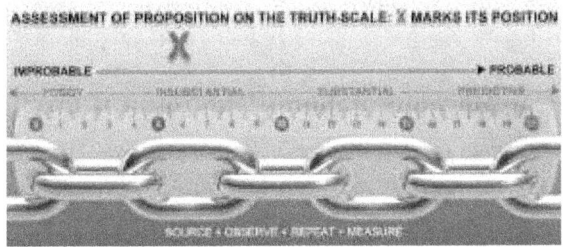

The authoritative figures in the community that hold this idea to be true have access to the same lacking chain of evidence as any of us and, so we can discount the authority of these convincing figures.

Another fall on our truth-scale for the idea of a soul, or spirit, or everlasting life-force.

Finally, what of the predictive ability of this claimed truth?

If the widely held idea of a soul, or spirit, or everlasting life-force was true then it should accurately predict the results of a recipient of the split-brain surgery, and yet it does not.

This places this claimed truth, very, very low on our truth-scale and therefore makes it highly improbable.

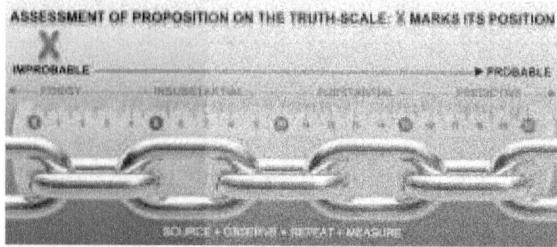

In conclusion, where should we place the proposition: Human reality begins and ends at the boundaries of our nervous systems. This proposition is quite reasonable and substantial given the many (independently verified) chains of evidence that support it.

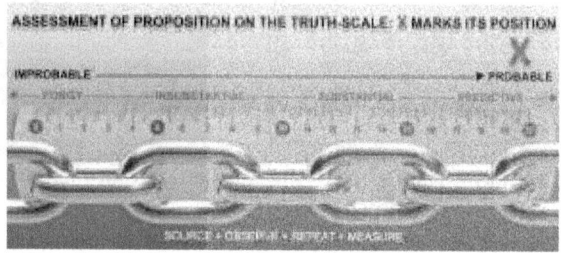

MENTAL ILLNESS AND THE SUPERNATURAL

Why is a belief in the supernatural akin to mental illness? You may be thinking that this position is too strong a claim or perhaps even a ridiculous one. Let us examine some ideas that may justify such a seemingly brash statement. We'll use our new system of defining truth again to place this proposition on our truth-scale.

Sigmund Freud (the famous Austrian neurologist and the father of psychoanalysis) proposed a key idea that has held up well and is now central to modern neural science. It is the notion that normal mental life and mental illness form a continuum. Mental illnesses, often represent exaggerated forms of normal mental processes.

If belief in the supernatural is akin to mental illness and if mental illness is the far-end of a continuum of normal human mental processes, then is belief in the supernatural inevitable.

Let us start the exploration of these notions by examining the concept of delusion.

What is the psychological definition of a delusional state?

An <u>improbable</u> and personal belief system or idea that is <u>not seen in a person's culture</u>.

The key to this definition are the terms: "improbable" and "not seen in a person's culture". Another way of thinking about this is the idea of popularly held truths.

Popular truths are often widely held cultural truths. As we have seen, widely held cultural truths may not be very reliable truths, but their ubiquitous presence in everyday discourse seems to somehow elevate their stature.

To explore this notion of delusion let us consider the following two stories:

A man visits his psychotherapist. During his visit the man says: "I hear voices every night before I go to bed."

The psychotherapist asks, "Is this all you experience?"

The man replies, "No, I can communicate with an invisible being and this being knows what I've done and what I'm thinking every second of every day."

The psychotherapist enquires, "How does this make you feel?"

"I feel warmth and well-being and close to my invisible friend."

Now, consider this story:

A man visits his spiritual advisor at his church. During his visit the man says, "I hear the Lord Jesus every night before I go to bed."

The spiritual advisor asks, "Is this all you experience?"

The man replies, "No, I can communicate with our Lord Jesus and he knows what I've done and what I'm thinking every second of every day."

The spiritual advisor enquires, "How does this make you feel?"

"I feel warmth and well-being and close to my savior."

There is an interesting linkage here, delusion seems to have a relationship with the concept of faith.

What is the definition of faith?

From the Oxford English Dictionary:

Faith
Noun
1. Complete trust or confidence in someone or something.
2. Strong belief in the doctrines of a religion, based on spiritual conviction rather than proof.

The second definition is the most relevant to our conversation here. Most humans alive today hold at least one truth founded on a strong belief in the doctrines of a religion, based on spiritual conviction rather than proof.

In the stories we have been considering above what is the difference between the two men?

The only difference is what is considered a popular idea and what is not. In both stories the men hold exactly the same belief, with the same lack of evidence to support that belief. However, the man in the church context is holding a popular and widely held cultural belief and the man in the psychotherapist's office is not in this same popular context.

It is quite reasonable that the psychotherapist in our story would diagnose his patient as delusional. His belief in an invisible friend is very improbable to say the least.

However, the man in the church context is seen as a faithful believer and not at all delusional. Yet both men hold the same supernatural belief.

Why is one man delusional and the other is not?

Because the church goer is surrounded by a popularly held idea, which is ubiquitous in his culture.

Let's continue our story and consider that holding widely held beliefs (that are not supported by chains of evidence) as true may have serious consequences.

After his visit to the psychotherapist our first man returns home and experiences a euphoric state while communing with his invisible friend. Our second man leaves his church, returns home, and experiences a euphoric state after communing with his savior. Both men have their experiences for a prolonged period. Most importantly, both men see and hear an invisible entity.

Consider for a moment that unlike schizophrenia, delusional disorder usually does not cause hallucinations. Individuals with delusions tend to retain organized thinking. They are also usually emotionally expressive, and report feelings that are consistent with their perception of what is going on in the world.

However, hallucinations are a hallmark of schizophrenia. Here is the psychological definition of the mental illness schizophrenia:

> A long-term mental disorder of a type involving a breakdown in the relation between thought, emotion, and behavior, leading to faulty perception, inappropriate actions and feelings, withdrawal from reality and personal relationships into fantasy and delusion, and a sense of mental fragmentation.

With this definition in mind let us continue the story of our two men.

Both men's experiences begin to lead them to faulty perceptions of reality. They both begin to feel inappropriate feelings and perform inappropriate actions. They both believe their invisible friend is real. They both begin to project their invisible friend into the reality they perceive around them.

Surely these men are hallucinating?

Is it possible that holding a belief in something that is not supported by a chain of evidence, could lead to delusion or aggravate the onset of schizophrenia?

This may be so if contradictory beliefs, ideas, or values are encountered by the sufferer. This phenomenon is known as cognitive dissonance.

Cognitive dissonance is not a mental illness, but rather a mental stress or discomfort experienced by an individual who holds two or more contradictory beliefs, ideas, or values at the same time.

Cognitive dissonance can precipitate anti-social or even criminal behaviors, or exacerbate mental illness. It can manifest itself when an individual performs an action that is contradictory to one or more beliefs, ideas, or values the individual holds to be true. It can also manifest when the individual is confronted by new information that conflicts with existing beliefs, ideas, or values.

When a person's collective truths, which are the foundation of their personal reality is challenged by facts or contrary ideas this can cause some individuals to perform unthinkable acts of violence. To explore this notion let us continue our story of the two men:

Eventually both men hear a message from their invisible friends. The message is: "Abortion is wrong and to right this wrong you must kill everyone who conducts abortions."

Both men understand that killing someone is wrong. Taking life is never justifiable. Also, they both must reconcile that the law of the land states the life of an unborn collection of cells <u>does not</u> take precedence over the life of a woman carrying these cells.

The first man returns to his psychotherapist explains his hallucination and the conflict he is experiencing between his supernatural directive and the law of the land.

He is given counseling and treatment to help him resolve these two conflicting messages. In this case cognitive dissonance is avoided by the treatment of delusion and/or schizophrenia.

The second man returns to his spiritual advisor who confirms that an unborn collection of cells has more right to life than the mother carrying them, and that the holy scripture confirms this idea. In this case cognitive dissonance is enflamed.

Here is a news quote by Polly Mosendz of Newsweek written on December ninth, 2015:

"Robert Lewis Dear shouted 'I am guilty,' Wednesday afternoon at a Colorado court hearing where he was charged with 179 felonies in connection with a November 27 shooting at a Colorado Springs Planned Parenthood.

Police officer Garrett Swasey, Army veteran Ke'Arre Marcell Stewart and mother Jennifer Markovsky died and nine people were injured when Dear, 57, opened fire at the facility.

Dear was arrested following a five-hour standoff with police. After his arrest, reports indicated he discussed 'baby parts' during police questioning—likely referring to videos released by a right-wing group earlier this year that claimed Planned Parenthood was involved in selling fetal tissue.

On Wednesday, Dear solidified suspicions that his opinion about abortion was a motive for his attack on the reproductive health clinic.

'I am a warrior for the babies,' Dear is relayed as yelling in court. 'Seal the truth. Kill the babies. That's what Planned Parenthood does.'

'You'll never know the amount of blood I saw in that place,' he added. CNN reported Dear made at least 17 loud remarks during Wednesday's hearing."

This news story illustrates the worst that can happen when cognitive dissonance is left untreated, or encouraged. But, what do we do when the beliefs that lead to this behavior are considered righteous; not mental illness, faith and not delusion?

We need the correct diagnosis and treatment of mental illness to reduce the instance of criminal and antisocial behavior. We also need to expose the risk that holding a belief in stories not based in chains of evidence can lead to cognitive dissonance and criminal anti-social behavior.

If belief in ideas not supported by a chain of evidence were innocuous and benign holding them would not be a risk to society. The tragedy above is the reason why condoning and promoting stories, not supported by chains of evidence, to the level of absolute truth is a very dangerous pastime.

Is there any way to immerse in fantasy without colliding with cognitive dissonance?

Yes, there is.

Consider the avid Star Trek fan as an example of immersing in fantasy in a healthy way. The popular television shows and movies have a large fan base. Regularly, many of these fans congregate at Star Trek conventions. It is quite common for these fans to dress in Star Trek costumes, to wear makeup, and to act out scenes from the popular show. Star Trek fans have even created new languages based on the alien civilizations from the show.

Why is it that Star Trek fans typically do not perform antisocial acts?

This is because the typical Star Trek fan knows that Star Trek is a fiction and is not a truth.

It is clear holding a fiction as a truth is very dangerous and can lead some individuals to extremely antisocial behaviors as we have seen in the abortion clinic example. But, indulging in fiction need not be an issue if we do not raise the fiction to the level of truth.

Sigmund Freud stated there seems to be no absolute or static human mental health. It is likely that all humans fall on a scale of mental health. It is quite possible that we all have a propensity towards schizophrenic or schizotypal personality disorder. It is also possible that we all have a propensity to believe in the supernatural, some more and some less. There may even be an evolutionary explanation for this human propensity.

Shamans arose in our hunter gatherer groups to satisfy the compelling human need to have answers to questions derived from our environment.

Individuals who were more susceptible to believing the Shaman's stories would cooperate better in the group, socialize, mate, and therefore pass on their genes. Those that were more resistant to the

Shaman's stories had less opportunity to thrive within the group and so were selected away over time.

Researchers have indeed recently identified a gene that increases the risk of schizophrenia in individuals.

After conducting studies in both mice and humans the researchers found a schizophrenia-risk gene called C4. They have a plausible theory as to how this gene may cause schizophrenia. This so-called schizophrenia gene appears to be involved in eliminating connections between neurons. This process is called "synaptic pruning", which, in humans happens naturally in our teen years.

The researchers speculated that it is this possible excessive or inappropriate "pruning" of neural connections that could lead to the development of schizophrenia. This would explain why schizophrenia symptoms often first appear during the teen years.

If this idea is so, it must be tempered by the notion of nature and nurture. Even if humans do have a gene that is directly linked with schizophrenia it does not ensure we will suffer from mental illness that could produce antisocial behavior, as the world we grew up in could temper this propensity.

This leads to the question would religion and belief in the supernatural exist if it were not passed on from parent to child?

Clearly, the girl in our crocodile story would not have held as true her belief that supernatural crocodiles live in the river if her father had not invented the story and convinced her of it.

Researchers, in cognition, religion, and theology at the Centre for Anthropology and Mind at Oxford University, have recently discovered that belief in the supernatural may fill the human need for finding meaning, sparing us from existential angst, while also supporting social organization.

These researchers who study the psychology and neuroscience of religion are helping to explain why such beliefs are so enduring.

They're finding that religion may, in fact, be a byproduct of the way our brains work, growing from cognitive tendencies to seek order from chaos, to anthropomorphize our environment, and to believe the world around us was created for our use.

This would explain why every group of humans found across the world has developed some form of supernatural belief system. Belief in the supernatural might be innate to the human condition and our realities. Of course, this does not mean we may not resist this tendency.

Whatever the reasons for the widespread belief in the supernatural, faith in the supernatural leads to self-delusion. This self-delusion is supported by, and aggravated by, inflexible systems of truths based on absolutes.

Why does holding supernatural stories to be true sometimes lead to antisocial behavior?

Because when the imaginary is believed as reality it causes paradoxical and contradictory thinking, which produces a mental conflict that in some cases amplifies mental illness, which can lead to antisocial behaviors.

It seems we all have a propensity to hold as true unsupported supernatural stories and this propensity is strongest in children.

The only way to break this cycle is to discontinue teaching children claimed truths that are not based on chains of evidence.

Is a belief in the supernatural akin to mental illness? It seems that this truth is highly probable.

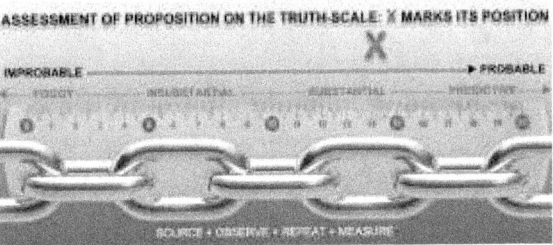

BIAS OR TRUTH?

One of the big problems we humans face is we tend to be more open to new information that confirms our previously held beliefs about reality and ourselves. We are far less excepting of chains of evidence that do not support our current realities, which of course are reinforced and supported by our collective truths.

We are much more accepting of truths that confirm our opinions and biases and are very likely to reject those that don't support our identity and world views. This phenomenon is known as confirmation bias, also called confirmatory bias or myside bias.

It is our tendency to search for, interpret, favor, and recall information in a way that confirms our already believed truths, while giving disproportionately less consideration to alternative truths and their associated realities. It is a type of cognitive bias and often causes a systematic error of our inductive reasoning capabilities. It can place a fog over otherwise clear chains of evidence.

People display confirmation bias when they gather or remember information selectively, or when they interpret it in a biased way. The effect is stronger for emotionally charged issues and for deeply entrenched beliefs.

Peter Cathcart Wason, who was a cognitive psychologist at the University College in London, conducted a series of experiments in the 1960s to demonstrate that people are indeed biased towards confirming their existing beliefs.

Another view of the phenomenon suggests that people show confirmation bias because they are hardheadedly assessing the costs of being wrong. This precludes them from investigating in a neutral way new evidence.

People tend to test ideas in a one-sided way, focusing on one possibility and ignoring alternatives. Explanations for confirmation bias include wishful thinking and our limited human capacity to process information.

Confirmation bias has also been put forward as an explanation of the following human psychological conditions:

Attitude polarization (when a disagreement becomes more extreme even though the different parties are exposed to the same evidence),

belief perseverance (when beliefs persist after the evidence for them is shown to be false),

the irrational primacy effect (a greater reliance on information encountered early in a series),

and illusory correlation (when people falsely perceive an association between two events or situations).

Confirmation biases contribute to overconfidence in personal beliefs and can maintain or strengthen beliefs in the face of contrary evidence. Poor decisions due to these biases can be found in all human contexts.

Given our tendency towards confirmatory bias we may all benefit from heightened skepticism, rigorous self-examination, and by questioning all simplistic or absolute truths we hold or are exposed to.

So where do we place the proposition on our truth-scale that: We humans are much more accepting of truths that confirm our current opinions and biases and are very likely to reject those that don't support our identity and world views. This is a highly probable truth.

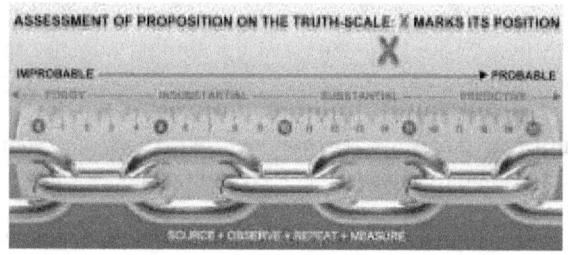

WHAT IS A LIE?

L ies are not the opposite of truths, which is the popularly held notion.
As we've discovered with our new system of measuring truth and our truth-scale: reliable truths are never binary, black or white, or absolute. A probable truth is a supple thing and something that can change over time when more information is discovered.

Our new way of understanding truth gives it the property of probability too. Some truths are very improbable, and others are highly probable.

Our truth-scale:

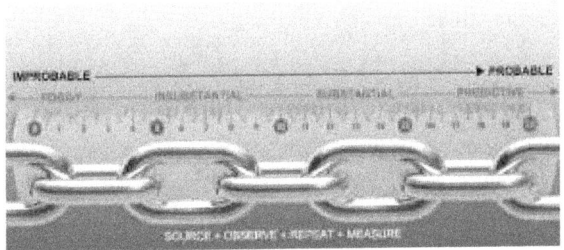

However, lies are different.

Here is the Oxford English Dictionary definition of the word lie:

Lie

Noun

1. An intentionally false statement.
2. Used with reference to a situation involving deception or founded on a mistaken impression.

Lies unlike truths are intentional misdirections often used for gain by the proclaimer of the lie. An example of a clear lie is the behavior and propaganda put forward by tobacco manufacturers.

Court hearings in the USA exposed that tobacco manufacturers knew about the addictive nature of their products and, yet they publicly denounced any such knowledge. They even funded pseudoscientific studies to keep the true nature of their product obscured. They did this to ensure sales and profits.

The popular view is that lies are always bad. However, this is not necessarily the case. Some lies may inadvertently lead to positive social results, just as some lies lead to very bad social consequences.

Take for example our crocodile story. In the end the father did purposefully deceive his daughter (lie). He deliberately (exaggerated) the crocodile threat, but his intentions were good from his perspective. He felt his (lie) was acceptable. He didn't want his daughter to be eaten by a crocodile.

The result of his lie was that his daughter was indeed not eaten by a crocodile. Therefore, one could argue the intention of this lie was for a good motive. However, the girl elevated his story to the level of an absolute truth and became convinced that the river was overrun by mystical crocodile creatures.

The father didn't set out to warp his daughter's reality, he wished only to protect her. However, if the girl went on to become a hunter of crocodiles her mistaken view of reality might lead her to the wrong conclusion. She might feel compelled to eradicate all crocodiles in all rivers. This would indeed be a bad, and unintentional consequence of the father's lie.

But, there is no escaping the fact that in some cases lies are used to intentionally harm others.

The tobacco companies cause millions of people to die unnecessary, slow, debilitating, and painful deaths for the sake of the tobacco company's profits. They deliberately withheld evidence of the addiction and cancer inducing properties of tobacco for their gain.

However bad or good lies are, they are not the opposite of truths. Lies are often intentional deceptions used to gain advantage over others. There is nothing gray about tobacco companies lying to maintain their profits.

Whether a lie is created with good intentions or deliberate bad intentions is something to consider when one is confronted by a lie. But in the end a lie seeks only to mislead.

Lies <u>are not</u> the opposite of truths. This seems a highly probable proposition.

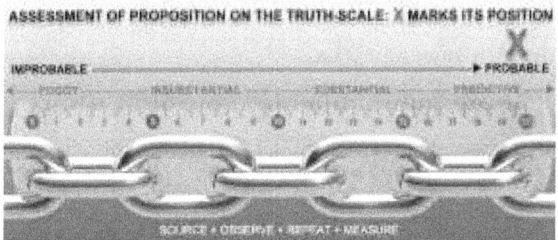

By David Millott

THE TRUE NATURE OF TRUTH

S o many things in our lives present themselves as truth. A shampoo advertiser claims that its product will make your hair shiny. A tobacco company insists that nicotine is not addictive. A drug company asserts you'll be happier if you take their drug. The Christians claim Jesus was God, lived, died, and rose from the dead. Muslims insist that Muhammad was the last prophet sent by God. The Hindu claim after biological death the soul or spirit can begin a new life in a new body. Judaism tells us that God created the Earth in six days and rested on the seventh. Australian Aboriginals explain that creation is the work of culture heroes who traveled across a formless land, creating sacred sites and significant places of interest in their travels. Uri Geller (Israeli illusionist) asserts he can bend spoons with his mind.

These are all things claimed to be true by their advocates often sincerely, but sometimes not.

So many claimed truths. So many people who believe something is really true. But, now we know how to determine what is a good truth and what is not.

James Randi (retired, famous stage magician, and scientific skeptic) has presented much evidence that bending spoons is a magician's trick. He has written many articles criticizing beliefs and claims regarding the paranormal. He has also demonstrated flaws in studies suggesting the existence of paranormal phenomena.

For example, in his Project Alpha he successfully planted two fake psychics in a privately funded psychic research experiment. Project Alpha demonstrated the shortcomings of many paranormal research projects even at the university level.

Most nonnative Australian's would find Australian Aboriginal's dreamtime stories untrue because their only evidence is hearsay. Most geologists and astrophysicists would find the Jewish creation story fallacious. Most Muslims would find the Hindu reincarnation story false. Most Christians would dispute the Muslims claim of Muhammad's connection with their God. Most of us would be skeptical of a drug company's claim that their drug could improve our happiness. Most addiction experts can produce evidence that nicotine is more addictive than heroin. Does your hair really become shinier after shampooing with any particular shampoo?

Now we can make an informed decision about what is true and what is not. More importantly we know how to test what we already believe to be true; is true. We now understand that truth is malleable, it is not binary or absolute. We know a reliable truth will be supported by chains of evidence and this evidence should always be: observable, repeatable, and measurable. We now know that we cannot claim to have an open mind if we are unwilling to change our minds based on new evidence.

Uri Geller has pledged that he uses only his mind to bend spoons. Australian Aboriginals see only their creation story as true. The faithful Rabbi insists God created the Earth in six days and will here no argument against his belief. The Hindu will defend their truth passionately, unwavering in their faith. Some Muslims will at times resort to physical violence to convince you of Muhammad's connection with God. Some Christians will kill patients and doctors at abortion clinics because of their belief in their holy stories. Tobacco companies ignore and are exempt of any responsibility of the deaths of some 20,000 people a month from tobacco related illnesses. Shampoo advertisers care only about sales and little about shinier hair.

We now know that if a truth is claimed to be absolute we should be highly skeptical about it and immediately put it to our truth tests.

What does it mean when truths contradict each other?

In most cases, proclaimers of truths are positive that their truth is true. And yet, someone will hold a contradictory absolute truth. But, how can two contradictory absolute truths both be true? If the Hindu reincarnation truth is contradicted by the Judeo-Christian-Islamic truth, how can they both be true?

Surely one, or both are not true?

The idea that contradictions do exist between truths is not a bad thing it is, in fact, a good thing. It gives us a profound insight into the true nature of truth. It is the first sign of a crack in the claim that truth is absolute, black and white, or binary.

Once we understand the true nature of truth human behaviors and events in the world begin to make sense. What seems impenetrable becomes clear.

The true nature of truth reveals that claims of absolute truth are imaginary. Assertions of absolute truth are replaced by levels of truth. This understanding causes the absolute perception of truth to vanish in a puff of logic. This small, imperceptible fracture in logic opens up to a vast world of understanding, knowledge, and wisdom. This tiny crack in logic can, and will set you free. It will enable you to ask the questions:

What is truth?

And why do I believe what I do?

From here the universe will unfold before your very eyes.

EPILOGUE

I n Ancient Greek philosophy the word Aletheia is the word for truth or disclosure. Aletheia has been translated as unclosedness, unconcealedness, or disclosure.

During the Roman era, Veritas, was the word for truth. Also, in Roman mythology, Veritas was the goddess of truth, a daughter of Saturn and the mother of Virtue. It was believed that she hid in the bottom of a holy well because she was so elusive. Her image is often shown as a young virgin dressed in white.

The idea of a sense of (something that is true) was first recorded in the mid fourteenth century. The idea appears in the kingdom of the West Saxons, which was an Anglo-Saxon kingdom in the south of Great Britain spanning from 519 CE until England was unified in the early tenth century. Then the word for truth was: triewð.

In the Mercian Kingdom that lasted from 600 to 900 CE the word for truth was: treowð. In this period, it meant: faith, faithfulness, fidelity, loyalty, veracity, quality of being true, pledge, or covenant.

The word truth, with the meaning of accuracy or correctness is found in English from the 1560s on. Interestingly, most other Indo-European languages of the time do not have a primary verb for (speak the truth), as a contrast to lie.

In modern times the word truth is most often used to mean being in accord with fact or reality, or fidelity to an original or standard.

Truth may also often be used in modern contexts to refer to an idea of (truth to self), or authenticity.

The concept of truth is discussed and debated in several contexts, including philosophy, art, and religion.

Many human activities depend upon the concept of truth, where its nature as a concept is assumed rather than being a subject of discussion. This idea of truth is used in most (but not all) of the sciences, law, journalism, and everyday life.

Commonly, truth is viewed as the correspondence of language or thought to an independent reality, in what is sometimes called the correspondence theory of truth.

The correspondence theory of truth states that the truth or falsity of a statement is determined only by how it relates to the world and whether it accurately describes or corresponds with that world. Correspondence theories claim that true beliefs and true statements correspond to the actual state of affairs.

Other philosophers take this meaning to be secondary and derivative. According to Martin Heidegger (German philosopher and a seminal thinker) the original meaning and essence of truth in Ancient Greece was unconcealment, or the revealing or bringing of what was previously hidden into the open, as indicated by the original Greek term for truth, Aletheia. In this view, the conception of truth as correctness is a later derivation from the concept's original essence, a development Heidegger traces to the Latin term Veritas.

C.S. Pierce (the American philosopher, logician, mathematician, and scientist) takes truth to have some manner of essential relation to human practices for inquiring into and discovering truth. Pierce himself held that truth is what human inquiry would find out on a matter, if our practice of inquiry were taken as far as it could profitably go. He said:

> The opinion which is fated to be ultimately agreed to by
> all who investigate, is what we mean by the truth.

Various theories and views of truth continue to be debated among scholars, philosophers, and theologians.

The method used to determine what is a truth is termed a criterion of truth. There are differing claims on such questions as what constitutes truth:

1. A truth-bearer is an entity that is said to be either true or false and nothing else;

2. How to define and identify truth;

3. The roles that faith-based and empirically based knowledge play;

4. And whether truth is subjective or objective, relative, or absolute.

Friedrich Nietzsche (the German philosopher, cultural critic, poet, and Latin and Greek scholar) famously suggested that an ancient metaphysical belief in the divinity of truth lies at the heart of, and has served as the foundation for, the entire subsequent Western intellectual tradition. He said:

> But you will have gathered what I am getting at, namely,
> that it is still a metaphysical faith on which our faith in
> science rests, that even we knowers of today, we godless
> anti-metaphysicians still take our fire too, from the flame lit
> by the thousand-year old faith, the Christian faith which was
> also Plato's faith, that God is Truth; that Truth is "Divine"
> ...

But I think we all now know, what the truth is about truth.

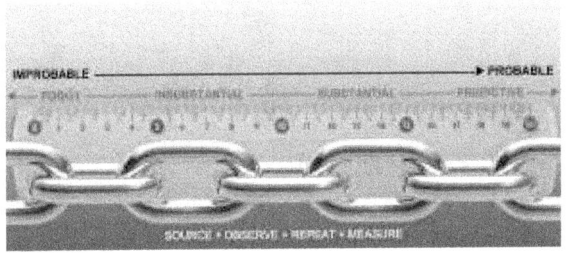

ABOUT THE AUTHOR

D avid Millett is a digital artist. He is an accomplished author, filmmaker, producer of paper books, and eBooks. He loves writing, videography, photography, filmmaking, travel, and his wife Julia.

From David Millett Publications:

Homo Cosmiens: Novel, Fiction, Sci-Fi, Drama

Femlet: Screenplay, Drama, Shakespeare, Hamlet

Anthropocene: Book, Science, Environmental Science

My Art, Volume One: Book of Artwork

The Creature: Novel, Fiction, Sci-Fi, Drama

Flying the Edge of America: Picture Book, USA Travel

Continental Drifting: Picture Book, World Travel

Tricky Nick Trickadee: Picture Book, Comedy

Table Talk: Short Film, Comedy

Knowland: Documentary Film, Environment

Autumn's Fall: Short Film, Drama

The Dog: Short Film, Drama

The Dialog: Short Film, Drama, Expressionism

The Root Cause: Documentary Film, Environment